Dream Affimals

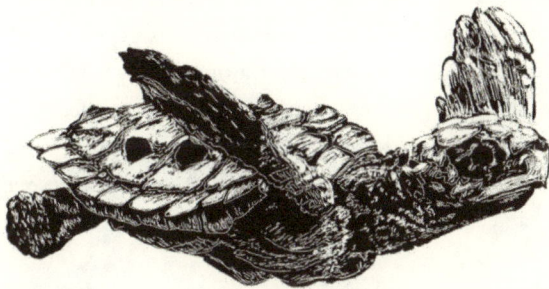

Dream Affimals

Affirmations + Animals

Inspiration to Fulfill
Your Wildest Dreams

Written and illustrated by
Elaine Miller Bond

SUNSTONE
PRESS

SANTA FE

Sunstone books may be purchased for educational, business, or sales promotional use.
For information please write: Special Markets Department, Sunstone Press,
P.O. Box 2321, Santa Fe, New Mexico 87504-2321.

Book and Cover design › Vicki Ahl
Body typeface › Jenson Pro
Printed on acid-free paper
∞

Library of Congress Cataloging-in-Publication Data

Bond, Elaine Miller.
  Dream affimals : affirmations + animals inspiration to fulfill your wildest dreams / by
Elaine Miller Bond.
    pages cm
  ISBN 978-0-86534-946-9 (softcover : alk. paper)
  1. Animal behavior. 2. Ethics. I. Title.
  QL751.B577 2013
  591.5--dc23
                              2013015883

WWW.SUNSTONEPRESS.COM
SUNSTONE PRESS / POST OFFICE BOX 2321 / SANTA FE, NM 87504-2321 / USA
(505) 988-4418 / ORDERS ONLY (800) 243-5644 / FAX (505) 988-1025

To all wild things in all wild places

Contents

Preface

I wonder what an animal might say, if it could offer us some words of advice. Nature's engineer, the American beaver, might suggest we "Still the waters." The humble earthworm, whose spaghetti body can drill into solid clay, would remind us, always, to "Stay grounded."

"So," people often ask me, "Which came first … the animal or the affirmation?"

"It would have to be the dream," I tell them. First I dream. Then, as if by magic, words flutter toward me, and silent messages emerge from the print left by slinking paws.

Some of the animals that winged or wiggled their way into this book are common to the global experience, and they offer insights into sharing (honeybees), strength (spiders), and why size, and number of feet, really don't matter (millipedes). Others hardly seem like wildlife at all: the chambered nautilus is a seashell most familiar to us in cross-section with the mathematical markings for a spiral, and coral is, in fact, both an unusual animal and the foundation of one of the world's richest types of ecosystems. All serve as wildlife ambassadors from lands as diverse as the American Southwest (peccary, kangaroo rat, bighorn sheep, to name a few), the far melting north (walrus, wolverine, Greenland white-fronted goose), Japan (Japanese snow monkey, red-crowned crane), China (giant panda), and Republic of the Congo (mandrill). So when we hear tragic news from Afghanistan, just maybe we will remember those snow leopards softly padding through high rugged mountains.

This book holds yet another underlying message—that predators need love, too. Meat-eaters are often sought after and killed when glimpsed near people or livestock, a situation they can hardly avoid in our shrinking world. Coyotes, for example, cause less human death than lightning, bee stings or rogue cows, yet about a half-million of these song dogs are killed each year, supported, in part, by tax dollars. The more furtive puma is a top-cat whose presence benefits all levels of the ecosystem, making it an "umbrella species" for biodiversity. Yet urban development is splitting puma country into fragments, and the only types of puma left to scratch out an existence east of the Mississippi River are the one hundred or so Florida panthers.

Indeed, animal writing is not all-soft as baby mice and owl chicks, when species near and far suffer from habitat loss, global warming, endangerment, poaching, and intentional persecution. I nearly broke down over the American bison, during that moment in my research when I glimpsed a photo of their skulls, stacked as high as a factory. And just as I began to feel that the bison's story was too intense for me to tell, I remembered the rush of watching their herds in the wild: bulls the size of small cars bedecked in woolly capes and bonnets; sloping-backed calves gamboling after their mothers. I remembered the grunts, bellows, and roars and how dust rising off clomping hooves clogged my nose. I remembered the love. Then I wrote.

This book is for everyone pricked by curiosity and revving with passion. Set your wishes soaring! And like the Greenland white-fronted goose, whose migration is an ascent up the world's second-largest icecap, may *Dream Affimals* help you "Set a course" to higher places.

Acknowledgments

Dream Animals is a book of community, made possible by scientists who pulled themselves away from saving salt marsh harvest mice or tracking desert bighorn sheep to share hard-earned knowledge with me. These pages contain but a glimpse into these experts' contributions toward understanding and protecting wildlife around the world. Karen L. Eckert (WIDECAST: sea turtles), Kostas Papafitsoros (University of Cambridge, England: sea turtles), Sandra L. MacPherson (U.S. Fish and Wildlife Service: sea turtles), Anthony D. Fox (National Environmental Research Institute of Denmark: Greenland white-fronted geese), Sarah Turner (University of Calgary, Canada: Japanese snow monkeys), Zara McDonald and Ally Nauer (Felidae Conservation Fund: wild cats), Camilla Fox and Gina Farr (Project Coyote), John L. Hoogland (University of Maryland: prairie dogs), Theodore Manno (Catalina Foothills High School: prairie dogs), Jan Stock and Sarah Haas (Bryce Canyon National Park: prairie wildlife), Paul Selden (University of Kansas: millipedes), Mary V. Price and Barbara A. Carlson (University of California: kangaroo rats), Ryan M. Moody (Dauphin Island Sea Lab: gobies), Reginald Barrett (University of California: wolverines), Deb Pople (Reef Teach, Australia: corals), Djoko T. Iskandar (Institut Teknologi Bandung, Indonesia: flying frogs), David Wake (University of California: flying frogs), George Beccaloni (The Natural History Museum, England: flying frogs), Michael Kaspari (University of Oklahoma: gliding ants), Stephen P. Yanoviak (University of Arkansas: gliding ants), James Estes (University of California: otters), Merav Ben-David (University of Wyoming:

otters), Howard Shellhammer (San Jose State University: salt marsh harvest mice), John Bradley and Doug Cordell (U.S. Fish and Wildlife Service: salt marsh harvest mice), Brooke Langston (Richardson Bay Audubon Center and Santuary: pickleweed), Mike Perlmutter (Bay Area Early Detection Network: pickleweed), Katie Yankula (Snow Leopard Trust), Peter Moller (Hunter College: electric eels), Horst O. Schwassmann (University of Florida: electric eels), Tim Clark (University of Hawaii: manta rays), Lacey Greene (University of Montana/Department of Fish and Game: bighorn sheep), Bighorn Institute, Darla White (Hawaii DLNR Division of Aquatic Resources: forereef zones), and Rich Bradley (Ohio State University: spiders): The animals and I thank you.

My utmost gratitude also extends to the following photographers who got to experience these animals for themselves and offered their inspiring images as references for illustration: Mick Sherington and the Australia National Parks and Wildlife Service (loggerhead sea turtle hatchlings), Jimmy Edmonds (Greenland white-fronted goose), Ingo Arndt/naturepl. com (baby snow monkey), Kenichi Nobusue (peacock), Helen Read and Paul Selden (millipedes), Paul Fernandez (giant panda), Gill Penney (giant panda cub), Ruth Savitz (African crested porcupine), Howard Shellhammer (salt marsh harvest mouse), Ryan M. Moody and Scott Beazley (frillfin goby), D. Gusman and D.T. Iskandar (Wallace's flying frog), Bill Stagnaro (California newt), Snow Leopard Conservancy, Alan Bond (honeybee, spider, American bison), Bruce Means (frilled lizard), David Fingerhut (zebra swallowtail butterfly), Peter Stubbs (palm cockatoo), Nigel Pye (barn owl), Jennifer J. Williams (baby peccary), David Mason (greater flamingo), Michael Bentley (chambered nautilus), Alexandre Shimoishi (red-crowned crane), and Kostas Papafitsoros (loggerhead sea turtle adult).

It does take a village (or shall I say "flock?"), and I happen to be blessed with talented, bighearted, nature-loving friends who helped immeasurably with *Dream Affimals.* For editing to referrals to hiding big, glossy spider books when they began scaring me, I would particularly like to thank Susan Gee Rumsey, William Harper, Marianne Betterly and her Thursday-night writers,

Alfred Miller and family, Emily Prud'homme, Janet Byron, Amy Foster, Pennie Opal Plant, Debby Kaspari, John Young, Cliff Moser, Kerry Drew, Lili DeBarbieri, Robert Cowart, Samantha Dugas, Riva Kahn Hallock, Lauren Kahn, Ona Russell, Vincent Ernano, Steve Patterson and flyingsquir-rels.com, Rusty Scalf, Linda Watanabe McFerrin and Left Coast Writers, WOM-BA, Meredith Maran, and Edie Meidav.

My editor at Sunstone Press, Jim Smith, is my ray of sunshine, who, from magnificent Santa Fe, turned *Dream Affimals* into a dream come true. If only my gratitude could gleam as brightly as his golden touch for books.

For seeing me through this project, as with everything in life, my deepest appreciation rests with my family. Mom, Dad, Alan: I love that you love these animals and believe in me.

Dream Affimals

Greenland White-Fronted Goose

Set a course

*M*igration is a climb up a mountain.

Its distant call reaches the Greenland white-fronted geese in springtime, when they feed in the peat bogs of Britain and Ireland. It beckons these birds to return to the marshes where they hatched. It moves them to fly as fast as 80 miles per hour over land, water, and ice. These geese migrate *over* the Greenland Ice Sheet.

Peak elevation along the birds' route … nearly two miles above sea level. Coldest conditions on the ice beneath them … 40 degrees below zero. Turning the thin air into power … their gift. Touching down on Greenland's remote western side brings a journey of more than 1,500 miles to a flapping close.

Weeks later, in this land of the midnight sun, fluffy brown goslings hatch in windswept marshes. Some of the goslings will stay with mom and dad for years. A few may spend their entire lives by their parents' sides, forgoing the chance to have chicks of their own and helping to care for other goslings when they hatch.

But none of these birds will settle here in Greenland. By autumn, white-fronted geese, young and old, feel that migratory call.

And back they fly, as families, toward the peat bogs. How the geese will navigate the return trip to their wintering grounds remains a mystery. Perhaps, like other migrating birds, Greenland white-fronted geese will find their bearings using the stars or magnetic fields. Most assuredly, though, they will face another tough, cold flight over the ice sheet. In the end, the journey back home is well worth the climb up any mountain.

The Greenland white-fronted goose (*Anser albifrons flavirostris*) is a distinctive race of white-fronted goose, named for the white feathers at the front of the head near the beak. All white-fronted geese migrate to northern tundras, but few amongst them undertake the treacherous, bill-chattering journey over the Greenland Ice Sheet, the second largest mass of ice on Earth.

Japanese Snow Monkey

Soak it all in

A Japanese snow monkey soaks in a natural hot spring, her teddy-bear arms flopped out upon the rocky edge. Her daughters—her life's companions—slide into the steamy pool and swim toward her, cradling babies of their own. The trials of winter melt away. The hungry, snowbound search for tree bark can wait. It all just rises into the sulfur-infused air. Three generations of monkeys: communing with water, mineral, and stone— warming to the core.

Bathing in natural hot springs, or *onsen*, is an ancient tradition in Japan. And in high, snow-covered mountains of the Nagano Prefecture, even the fluffiest of monkeys is happy to carry on the tradition.

Snow monkeys pass along other traditions as well, which vary from troop to troop, region to region. Some troops wash their food, others handle stones together, yet others roll snowballs for the joy of it. No matter the troop, all such traditions share one thing in common: they begin when a monkey gives something new a try.

Splish! Splash! Kurplunk!

One after the next, snow monkey infants take high, belly flopping leaps into the hot spring, temporarily leaving their moms to go play. Whereas most types of monkey avoid the water, these fluffy primates start off life with a splash, using swim-time to bond with one another, get into tree-climbing form, and just have fun. Their moms wait nearby with open arms. But even a snow monkey baby knows: some of life's moments are for soaking up the warmth; others are for taking the plunge.

Japanese macaques, or Japanese snow monkeys (*Macaca fuscata*), are the world's northernmost primates besides humans, tolerating a colder climate than all other monkeys. They typically sleep in the trees but will also take their rest on the ground, huddling in groups for warmth. These monkeys are endemic to (uniquely found in) Japan, where they are protected. Japanese snow monkeys were previously listed as endangered and are extinct from Tane Island.

Peacock

Be dazzling

May-awe. May-awe.

Banyan trees of India ring with metallic calls before the peacock glides down into the grasses, one of the world's most spectacular birds taking the likeness of a flying dragon. He is, in fact, known as an Indian (or blue) peafowl, a type of pheasant. The female is called a peahen, which, upon selecting the most dazzling peacock, will soon face a new responsibility at night in the trees: keeping her peachicks safely tucked under her wings.

Like the peahen, the peacock has a short tail. He also boasts a train of long covert feathers, which he opens into a glimmering iridescent fan. Sir Isaac Newton described this light effect as "structural color," because microscopic barb-shaped structures—rather than pigments—make the feathers shine like opal fire.

To a peahen, however, the peacock's shrill-sounding song may be equally dazzling as his feathers aflame. Most eager to flaunt his vocal prowess before the monsoons, he calls for both peahens and his fellow fancy brothers, the sound of hands clapping and even the thunder itself. Such is the beauty of showing off all you've got. The rains just may come pouring down.

Indian peafowl (*Pavo cristatus*) are native to seven South Asian countries and have been crowned the national bird of India. The peacock's glamorous train can account for nearly two-thirds of his total body length, and a male whose fanned feathers display more eyespots, or *ocelli*, seems to attract more females. He sheds his train in January (the off season), but by the summer monsoons (the mating season), he is back in full color and full swing.

Puma

Get more pounce to the ounce

A deer grazes in a meadow. But like a puff of smoke, a puma curls in and out of the rocks, undetected. Crouching low in the grass, the cat moves closer in slow motion, head stretched forward, legs wound tight, eyes fixed on the animal grazing within the pounce zone. Timing needs to be paw-perfect. And when the puma finally springs into action, its whiskers jut forward like a brush of feelers. Its paw pads touch the deer's fur, using the direction of hairs as a guide to the head and neck. The hunt ends with just one bite.

Deer are the puma's top choice, but with the ability to leap 20 or more feet, a hungry cat often gets what it wants. It can snag the world's second-fastest runner—the pronghorn—and overtake prey 500 pounds heavier than its own body, then drag it off to a private spot.

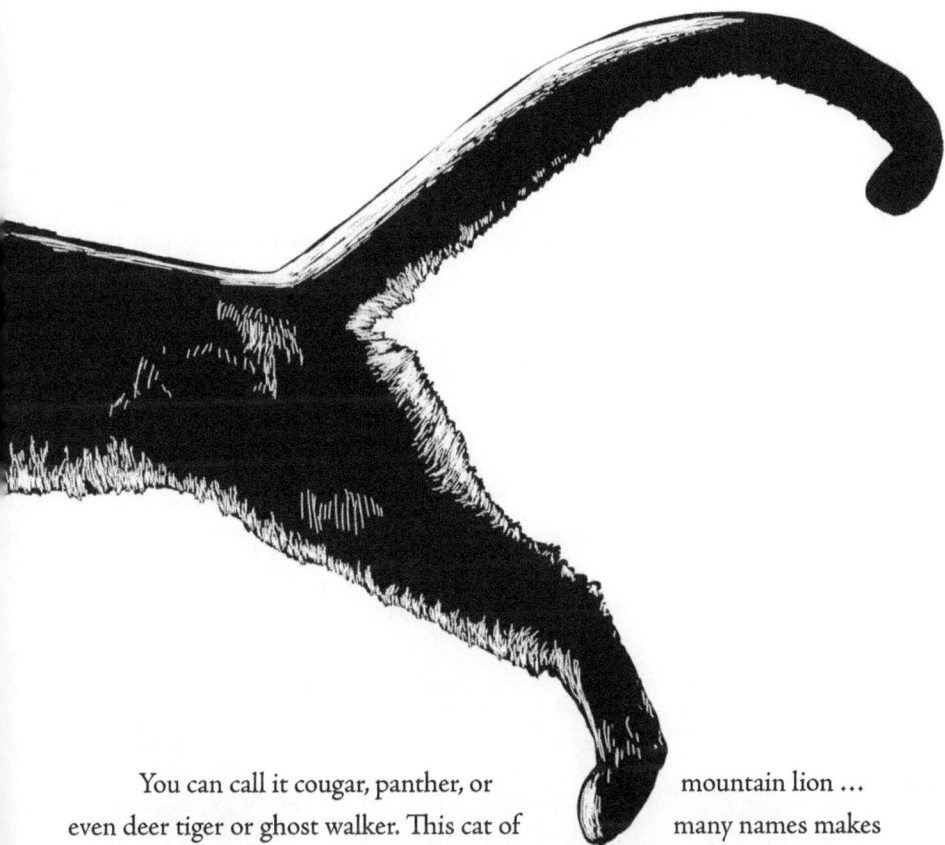

You can call it cougar, panther, or mountain lion … even deer tiger or ghost walker. This cat of many names makes its airy, stealthy way over much of the New World, being the most widespread mammal besides humans. Yet it's as elusive as ether, mostly seen after the fact, by its tracks. It's the power to keep your eyes on what you want and go capture it for yourself. It's the independent spirit, the pioneering spirit, the spirit of the wild Americas.

Lacking stripes or spots, the puma (*Puma concolor*)—or the puma "of one color"—is amongst the few plain-colored wild cats. It prowls and pounces from British Columbia to Patagonia and can adapt to habitats as diverse as deserts, mountaintops, and swamps. The Florida panther, an isolated subpopulation of the puma, is endangered.

Coyote

Sing your song

Golden yellow blossoms of the prickly pear cactus open with the sunrise, turning this needly plant into a desert beacon. And equally bright, equally sunny is the song of the Desert Family.

Bee sings to Pack Rat by pollinating the prickly pear—juicy enticement to a twitchy little nose.

Pack Rat sings to Rattlesnake by giving up its bric-a-brac den.

Rattlesnake sings to Bison when it steps a bit too close.

Bison sings to Prairie Dog with every chomp that keeps the grasses whisker-high.

Prairie Dog sings to Burrowing Owl in chittering alerts to danger: flying, slinking, slithering.

Burrowing Owl sings to Coyote by bobbing up and down on its mound—the metronome for the desert's beat.

Coyote picks up the beat and sings it back out to its mother, father, sisters, brothers, elders, cousins, babies.

In summer—the season of rodents—Coyote sings for togetherness after the lone hunt. In winter—the season of great herds—it sings in cheerful choirs that share the spoils. "Woo-oo-wow" is Coyote's greeting song, modulating high and low, over hill and swale, to connect its family of native song dogs across the Americas.

Sing, Coyote, sing …

Coyote takes a deep desert breath, throws back its head, and releases every howl, bark, yip, huff, and scream. Its soaring voice is part cat and dog, part fox and wolf, part bison and eagle, whose wings carry its music up and up in celebration of the rising midnight star, and lasting, like its loneliness and laughter, until the blooming of the sun.

One coyote (*Canis latrans*) eats about five rodents per day, plus insects, helping to keep the ecosystem in balance. This versatile canine can adapt to cities, and with urban expansion and the widespread persecution of wolves, the coyote has expanded its range beyond the arid lands and plains of the United States, Canada, and Mexico. Its song can now be heard as far north as Alaska and as far south as Panama.

Millipede

Jump in with both, both, both, both feet

Ten thousand species of millipede are known to pitter-patter about the Earth's moist places, and many more are waiting to be discovered. Some millipedes are tiny; others spiny. Some roll into a ball; others glow. Yet all millipedes have segmented bodies—and four legs per normal body segment—which gives these animals the zippery undersides for clinging, burrowing, or scrambling through the loose and rotted places of the forest floor.

Most millipedes, from the giant African millipede (*Archispirostreptus gigas*) to the so-called shocking pink dragon millipede (*Desmoxytes purpurosea*), have no more than three hundred legs in total. The world's record rests with the *Illacme plenipes* millipede of Central California. Counted under a microscope, one female was found to have 750 legs on her one-inch squiggle of a body. Yet even she fell short in representing the millipede name: "thousand-footed" in Latin.

But don't tell a millipede that! What really counts is that we feel a thousand feet long ... all millipedes being equal.

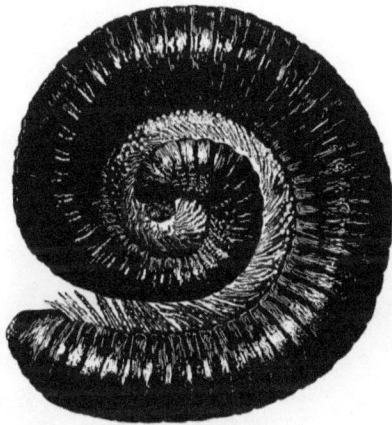

Millipedes (class *Diplopoda*) are the first animals, besides microbes, known to dwell on land rather than in the sea. A bus driver in Scotland with a penchant for paleontology discovered the oldest millipede fossil to date: 423 millions years old, from the Silurian Period. Centipedes, often mistaken for millipedes, have two legs per body segment instead of four. Centipedes are also quick predators, feeding primarily on bugs and worms. Millipedes are vegetarians that take their time.

Giant Panda

See beyond black and white

To a panda, a bamboo thicket is an easy chair. She reclines into it, her hind legs flopped open, belly hanging out, Mouseketeer ears rising and falling with every munch of the leafy cane. But in these clouded Chinese mountains, even precious bamboo is not enough for a panda. Soon this female will look for a dawn redwood, or some other grand tree, to use its hollow trunk as her den. This giant panda has a cub on the way.

The baby panda is born, tiny and pink and fitting comfortably into its mother's big warm paws. She will cuddle it continuously for weeks. Later, when the bears venture out together, mom will eat bamboo with one paw, snuggle her cub with the other. Then to the den they return, back to sleep, mama and baby curled up side-by-side, as one.

By several months of age, the cub starts taking time alone, playing up in the trees in its roly-poly panda way. Despite this new-found independence, the floppy young bear will spend eighteen more months cuddling and exploring with its mother before striking out on its own. Boy pandas stay close by. Girl pandas are especially adventuresome female mammals, sometimes journeying far away, to their own distant grand trees in their own distant bamboo forests.

Many have wondered why the giant panda evolved from a carnivore into an herbivore; its teeth can crush either cane or bone. They have also wondered why the bear's coat is black and white—colors that seem to blend with snow in winter and look bold in spring, when pandas search for a mate.

Carnivore. Herbivore. Black. White. None of this matters to mountains in the clouds. Here, what truly makes the panda a "giant" among bears is the tightness of its hugs.

Fossils of giant pandas (*Ailuropoda melanoleuca*) have been unearthed in Burma and Vietnam, and pandas ranged broadly across the Chinese highlands as recently as the 19th century. Now they live in small, fragmented populations, confined by areas of agriculture and logging. Bamboo, the bears' food source, undergoes natural cycles of mass die-off, posing further threat to their survival. Giant pandas are endangered.

Porcupine

Get to the point

*B*ut use the truth to tickle, not prickle.

Porcupines, or "quill pigs," are large, often bulky rodents.
Porcupines in the Americas (family *Erethizontidae*) dwell
in trees; some even have prehensile (grasping) tails.
Porcupines in Africa, Asia, and Europe (family
Hystricidae) are highly adaptable and can
dig burrows for shelter. Some swim.
Needle-sharp quills are modified guard
hairs, standing as high off the back
as twelve inches, as in the African
crested porcupine (shown here).
This species also uses
hollow, open-tipped
quills in its tail
as a rattle.

Kangaroo Rat

Let your energy spring forth

Kangaroo rats are hopping with excitement, as tonight in the starlight, seeds have freshly fallen upon the drylands of the American West. Desert poppy. Buckwheat. Yellow flowering lotus. Panic grass. These are just some of the plants that produce big, round, come-to-me seeds that the kangaroo rats scoop up with delicate little hands and place in the grocery bags of their fur-lined cheeks.

Each kangaroo rat scrambles to bury (hide!) most of its seeds in hundreds upon hundreds of tiny pits, which serve as caches for later use. Incredibly, the kangaroo rat manages to remember where every single pit-cache is located. (If only the kangaroo rats knew where we placed our car keys!)

Like a spring-loaded puppy, a male kangaroo rat goes for a roll in the dirt. Regular dust bathing keep his fur in top shape and rubs his oily scent around his territory. But he must do more to protect he seeds. The first quenching rain is coming; it will soak the land and the seeds, releasing a wonderful earthy aroma which unfortunately for kangaroo rats, reveals where their seeds are hidden.

We'd better do something!

Come nighttime, the desert bounces in a frenzy of springboard feet and pom-pom tails. Though working frantically, these rodent masters of memory must now keep track of *three* things: where they originally hid their seeds, where they moved their seeds, and where they stashed their friends' seeds that they dug up and stole.

Despite their bouncing ambition, however, the kangaroo rats' big, dark eyes and soft, balloony cheeks turn out to be much bigger than their stomachs. So they "give back" during the spring rains, when most of their seed collection remains buried underground, unnibbled, and sprouting into little blooming pockets—renewing the desert itself, and giving new life to birds, bugs, and of course, more cheeky, springing bundles of energy.

Kangaroo rats (*genus Dipodomys*) live in Death Valley and other hot, dry places in the American Southwest and Mexico. Yet they seldom drink water, relying, instead, on their ability to extract moisture from seeds and other food they eat. Water vapor in the air also condenses inside their long nasal passages—the rodent equivalent of a dog's cool, wet nose. The endangered Stephen's kangaroo rat and seven additional kangaroo rat species are included on the international list of threatened species, primarily due to urban and agricultural expansion into their habitats. The desert kangaroo rat (*Dipodomys deserti*) is shown here.

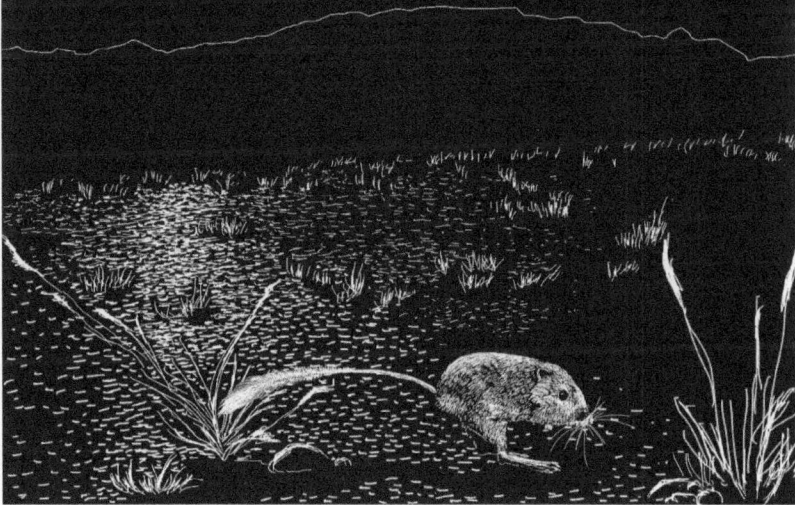

Frillfin Goby

Think on a grand scale

The frillfin goby is a small tide pool fish. When danger approaches, the fish makes grand leaps from one pool to the next, eventually out to sea.

But how does the goby jump so accurately? Maybe it draws a map of the pools when they flood together at high tide. Maybe it keeps track of its route all day. Or just maybe it does something we haven't yet imagined.

Way to go, Little Fish—thinking outside the pool. A whole ocean is waiting.

Gobies form a big family of mostly little fishes, some less than a half-inch long. As in other gobies, the pelvic fins of the frillfin goby (*Bathygobius soporator*) are fused together like a suction cup that can attach to rocks and other surfaces. This fish seeks shelter in shallow tidal zones of the United States, Central and South America, Europe, and Africa.

Blue Whale

Love big

Peaceful, powerful, more than two buses long, the female blue whale is the largest animal the Earth has ever known. Imagine how well our world would run if we all had her heart of half a ton.

Blue whales (*Balaenoptera musculus*) can reach 100 feet from head to tail and tip the scales at 160 tons. Females are slightly larger than males, and they go for months without feeding—and lose about 50 tons of bodyweight—as they nurse and bond with their calves. When not breeding or caring for young, blue whales lead rather solitary lives, and they can emit low-frequency sounds that carry for thousands of miles across ocean basins. These gentle giants live in all the world's oceans and find the best, krill-rich feeding grounds in the coldest waters to the north and south. Commerical whaling of the 19th and 20th centuries depleted the blue whales' global population. Today, they are endangered.

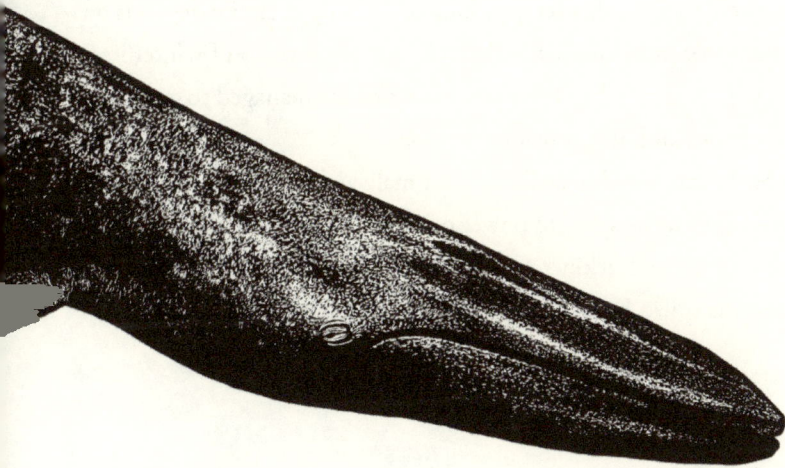

Wolverine

Keep your fighting spirit

A wolverine walks low through the snow, his tracks meandering wider than any frozen river. Wandering both day and night, he steals chances to lay his head down on the rocks and snow. But never for too long. In the cold northern world of the wolverine, opportunity may lie just beyond the next twist or turn.

The wolverine's snowshoe paws enable his search to press on under any condition, even a morning blizzard, when, somewhere in the same Norwegian forest, an elderly reindeer looses pace with its herd. Its hooves, like weights, sink into the fresh powder. The last thing this once-mighty animal will ever see is a furry silver mask with jaws.

This large kill is the promise of hunger melting away for the next three weeks, so the wolverine digs a burrow to hunker down beside it. Lacking speed, stalking skills, or even a tendency to stay quiet, the wolverine must fight extra hard for his food. And he will fight even harder to keep it. Any wolves, lynxes, or bears that dare challenge him would soon discover that strength can come in small, short-legged packages.

More often, though, this scrappy king of the weasel family depends on his foes for their frozen leftovers. He also relies on the severity of winter itself, a life lost helping to sustain his. Even as a young kit, he managed to get by on a meager diet: bones and hide provided by mom.

But now he has only himself. The next male wolverine is hundreds of miles away, giving him the space to power on through the snow and persist on what little it gives up. Luckily tonight his belly is full. He can lay his head down to renew his spirit. For tomorrow will be another fight.

Gulo, taken from the wolverine's Latin name (*Gulo gulo*), means "glutton," though this animal takes only what it needs and wastes nothing. It is very territorial and requires vast amounts of space—free from people and usually other wolverines—living in low densities in boreal zones across the northern hemisphere.

Mandrill

Be the rainbow

*R*ise and shine, mandrills!

High in the branches, a group of one hundred mandrills wakes, yawns, stretches long, furry, olive-green arms. A young male mandrill makes his way to the ground first, but rainforest trees make wobbly staircases. So he parachutes off a tree limb, snatches a vine in midair, and swings like a long-faced pendulum before he drops down. A mama mandrill makes branch-by-branch leaps toward the forest floor, her baby clinging tightly to her chest. The toddlers—all spindly arms and legs—trickle down a branch of their own, one going head first, one bottom first, one just sort of rolling.

By tonight, these monkeys will retire back up in the treetops. But today, they will cover a lot of territory, filling the sub-Saharan rainforest with flower-bright color and grunting noise as they go.

The dominant male mandrill—the world's biggest monkey—slowly clambers down from the canopy by hugging the tree trunks. He prefers to keep his feet on the ground, walking proudly as a lion and sporting the yellow beard and puffed crest of leadership. His big, round, eye-catching rump is a festival of fuchsia, violet, red, white, and yellow, which gets brighter as he gets excited. And there is a lot to get excited about, when the females are attracted to this largest, most colorful monkey.

The mandrills stop for a break, and the alpha male leaves the group to sit atop a grassy mound. Like a primate version of the Great Sphinx, he keeps guard over his family as the older groom the younger, mothers nurse their babies, and tots do their tumbling. But one monkey from the bunch, an infant, decides she would rather go play with dad. When she moves in close to the big male mandrill, he bears huge fangs.

No worries. Even the youngest of monkeys can recognize a mandrill's smile. For in Equatorial West Africa, and all around, the most amazing rainbows are found right on the ground.

The mandrill (*Mandrillus sphinx*) is native to four African countries: Cameroon, Equatorial Guinea, Gabon, and Republic of the Congo. This species is considered vulnerable due to commercial bushmeat hunting and loss of rainforest habitat. Its closest relative, the less colorful but equally handsome drill (*Mandrillus leucophaeus*), is endangered.

Coral

Let yourself bloom

Corals look like exquisite living bouquets in the sea, as vibrant as tropical fishes, solid and glimmering like gems. So why are corals in this wildlife book? They are animals.

Coral animals are small, tentacled polyps, much like sea anemones. And they slowly build homes of rock in shapes as fanciful as branches, vases, and even elaborate craggy towers over 100 feet tall. But corals cannot do it alone. Inside their soft bodies live tiny plants (algae), which give them energy. So corals possess a beauty that far surpasses their looks—that of Animal, Plant, and Mineral working together in harmony.

The Great Barrier Reef is the world's largest living structure, and here, corals carry out one of the most spectacular phenomena in the Animal Kingdom: they "bloom." The coral bloom is an event of synchronous mass-spawning, usually occurring once a year, a few nights after the full moon in November. On these nights only, most hard corals release flurries of gumball-pink bubbles, called *gametes*, each containing the material for reproduction. The *gametes* float to the surface, like a blizzard in reverse, forming pink slicks across the water so immense that they can be viewed from Space.

Underwater, the coral bloom may look like a freshly shaken snow globe, but these nights are highly choreographed. Corals across the Great Barrier Reef—an area the size of Japan—manage to coordinate long distance with one another such that one type of coral spawns, then another type spawns, until 150 coral species have sent their bubbly gametes adrift into the summer seas. Gametes will surface in synch with their own kind, mix amongst the waves, and create new baby coral animals.

The water on bloom nights is so pink and effervescent, rejuvenation so profuse, that deep-water fish swim in for a feast, and reef fish already in the neighborhood stay awake for late-night snacking. The flurry of life overwhelms the appetites of predators. So parrotfish, damselfish, even sea stars and clams use this time to send their own young out into the wide watery world.

A crab pinches its way up a staghorn coral, as if it were a tree, then dangles from one claw, snipping and snapping at delicious pink fizzies as they float by. Swaying, blooming, bubbling ... renewal always lifts you up.

Coral reefs cover a very small proportion of the seafloor—a fraction of one percent—yet they serve as home for about one-third of the world's marine fishes. Corals (class *Anthozoa*) live in shallow tropical waters, where sunlight can reach the tiny plants inside their bodies. Bleaching occurs when corals expel these plants and die, and with the warming of ocean waters, such events are occuring on a new, distressing scale.

Wallace's Flying Frog

Catch some air

Why keep your feet on the ground when you can use them to soar?

Wallace's flying frogs have parachutes for feet and webbing along their bodies, giving them the loft to glide in 50-foot bounds through the rainforests of Borneo. This is the high life, sailing from tree to tree, branch to branch, while searching for the next meal and weightlessly avoiding becoming one.

Even the tadpoles of flying frogs take a pass through the air. Their metamorphosis begins when the mother frog lays her eggs in a foamy nest, which she beats to a froth with her leg and hangs strategically from a branch. Her plan takes full effect when eggs grow into tadpoles—too heavy for the foam—and freefall through the air. They land with a spash in a temporary pool on the rainforest floor, a pool, in fact, that a wild pig had dug out. What was once a pig's wallow now becomes the murky place where tadpoles swim and slosh their way to adulthood.

Having traded their long tadpole tails for long webbed fingers, young flying frogs spring up into the trees, into the magical realm of sleeping sun bears and swinging orangutans, to spend their lives amongst the leaves and the air. But they are just part of Borneo's friendly skies; a whole menagerie goes gliding and parachuting up there. Flying dragons are real-life lizards that spread their winglike flaps and glide. Flying geckos get their lift on feet like flattened stars and tails like ferns. Flying squirrels and flying lemurs (which are not true lemurs, but speckled relatives of primates) are the fluff of the air, soaring sometimes hundreds of feet around trees and down hillsides. Flying snakes actually slither through the sky.

Whether these unlikely lofty creatures are the puffy masters of controlled falling or simply sailing to the next place in life, they all teach us to keep looking up. For it always saves energy and spreads beauty to take the high sky road.

Wallace's flying frogs (*Rhacophorus nigropalmatus*) are large treefrogs named after Alfred Russel Wallace (1823–1913), famed English naturalist and explorer. These frogs live in Borneo, Sumatra, and Peninsular Malaysia, usually staying high in the trees and out of sight, except when they descend for breeding. Sumatran rhinoceroses (*Dicerorhinus sumatrensis*), also known as hairy rhinoceroses, were once an important source of dugouts for growing tadpoles; today these rhinos are critically endangered and extinct from most of their original range. Habitat for flying frogs and other rainforest fauna, particularly in Sumatra and Borneo, is rapidly being cleared to make room for palm-oil plantations.

Eurasian Otter

Play around

Otter's advice for life:
Find a horse-chestnut seed, and use it for soccer practice.
Float downstream, spiraling and chirping in the arms of your mate.
Belly slide into a river, over and over and over.
Be a cub and play!

Continents of slinky otters know …
The best way is usually the fun way.

Also called common otters, Eurasian otters (*Lutra lutra*) have webbed feet for swimming and sharp claws for grasping eels and other slippery prey. They are the world's most widespread otters, sliding into streams of Europe, Asia, and North Africa and adapting their frolicking lifestyle to some coastlines. Eurasian otters are internationally recognized as near-threatened due to pollution and other anthropogenic (human-caused) pressures.

Walrus

Chill out

There is one place a dominant walrus wants to be—near the female group—and he will move Heaven, Earth, and Blubber to get there. He heaves his 1.5-ton body along the floating pack ice, prodding other bulls with his tusks and thrusting them aside. When he finally reaches the group, he flops onto his flank and takes a nap. Nothing beats basking near the females.

Mother walruses in the area hardly notice him. Their world revolves around calves, and they form a bond with their young for at least two years—longer than any other member of the *pinniped* (walrus, seal, sea lion) family. During this time, the attentive mother will nurse her calf, though a young walrus can eat on its own by slurping meat out of shellfish with its short, floppy face. She will also give her calf back-rides in the water, though a baby walrus takes to swimming like a fat, wrinkly fish. When the mood turns playful, the mother walrus lifts her calf up out of the sea with a tight embrace of her flippers.

Another walrus bull forks the ice with his tusks, hauls his body out onto a floe, and trudges toward the females. Today this younger male has a special reason for approaching the group. And he is not alone: he has pulled an orphaned calf out of the Bering Sea, its weak barking voice moving many walruses to lumber in close. Rescued by a bull and embraced by the herd, this little bundle of whiskers and flippers now gets its chance to bond with a new mom, who may someday, have a reason to hold her flippers high.

The slanted northern sun has now reached its maximum, and a few walrus bulls have fallen asleep in the water, floating by their inflated necks like huge, scarred, leathery bells. The other walruses are hauled out on the pack ice, piling atop one another for a snooze. And as the ice begins to drift, so do the calves, on their warm, breathing mattresses of blubber and tusks, off to walrus dreamland.

Walruses (*Odobenus rosmarus*) prefer the company of others, and they will gather, sometimes by the hundreds, on sea ice in a discontinuous ring around the Arctic and near-Arctic. Both males and females have long tusks, which they use to fend off polar bears, haul out onto the ice, and in the case of large bulls, fight for mates. While displaying for mates, bulls produce an elaborate array of bell-like tones, clicking noises, and other enticements from a nearby place in the sea. Walrus populations hit their lowest mark in the 1950s after decades of commerical hunting.

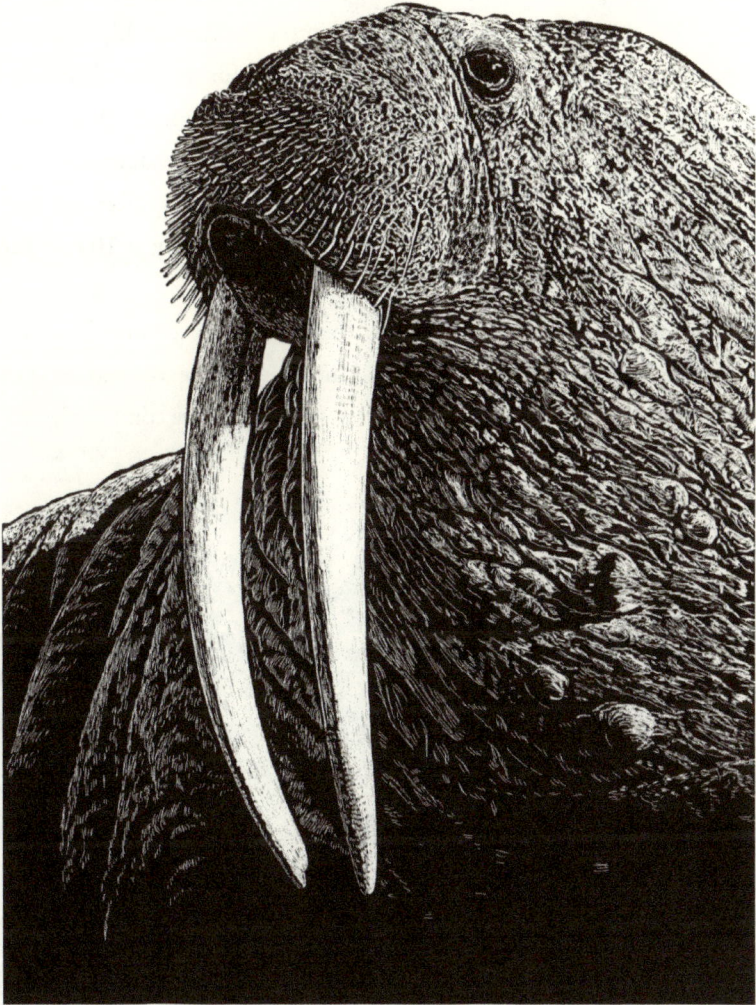

Earthworm

Stay grounded

Above us, the atmosphere is like the Earth's breath. Around us, the hydrosphere is every last drop that gives us a blue planet. And right beneath our feet, the drilosphere is the richest layer of soil—engineered by earthworms—living out their precious, wiggly lives by following their gut.

For the human gardener, digging into hard clay usually means hours on the knees and calluses on the hands. But not for this noodly creature. It's hydraulic! The earthworm builds fluid pressure in its body by pumping its hearts. (Usually they have ten!) This pressure then enables the animal to shoot forward and take a bite of earth. The tail contracts after every mouthful, and bite by bite, the stringy animal may tunnel as deeply as 10 feet into the ground.

The more the earthworm eats, the better, for everyone. This limp little chemist can munch into soil and into decaying leaves and meld them both together. The results are called *casts*, soils that are completely new and

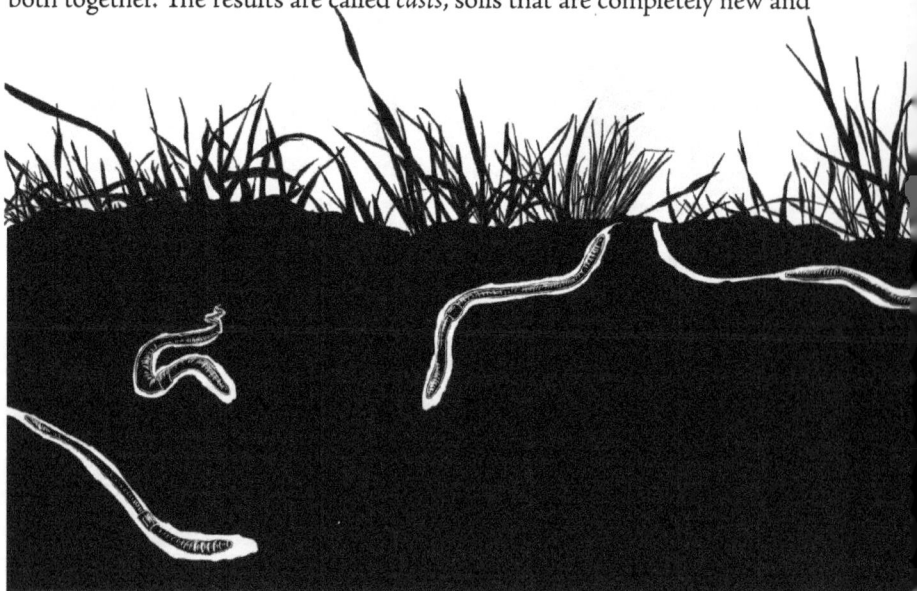

especially nutritious. The earthworm also provides living microbes from its gut and the slime of its trails, life support for soil and the roots of plants and trees.

These odd animals may seem like little more than spaghetti that squirms. They even lack eyes. But one bite at a time, they created the soils that fossilized the bones of dinosaurs and concealed fortunes in Roman coins. And quietly, within the solid darkness that supports us, they eternalize the past and ground the Earth.

Earthworms (from *Oligochaeta*) live in the soils and muds of every continent except for Antarctica. Some are aquatic. Famed English naturalist Charles Darwin observed and performed studies with earthworms for decades, dedicating his last book to these humble creatures and their important work.

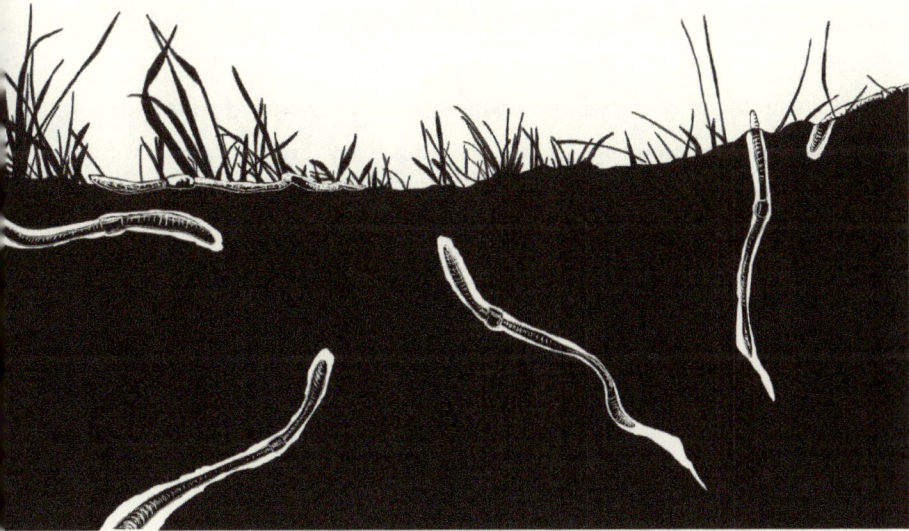

Salt Marsh Harvest Mouse

Eek the most out of life

Many mice view a grassland as a welcoming seed-filled sea. Salt marsh harvest mice, however, prefer seas of a different kind: real ones. These teeny furry creatures live in the wetlands along San Francisco Bay—a bustling urban center where the vast majority of original wetlands have vanished.

When the tide comes in, salt marsh harvest mice scurry to the higher, drier ground of the backmarsh (where it still exists). During the highest tides, these adaptable rodents climb up a coyote bush or other tall plant and swiftly mouse-paddle across the water to the next plant. But mice only swim as a last resort. Their usual m.o. ("mouse's operandi") is to keep undercover, slipping safely through the thick, dim matting low in the marsh, beyond the eyes of hawks and herons.

Great fans of over-seasoning, salt marsh harvest mice eat briny wetland plants as if they were cocktail pickles. They also have the extraordinary ability to drink saltwater. Some of these special harvest mice actually prefer saltwater, leaving the fancy California spring water for other mice. Life is damp and brief, about nine months long. And a mother mouse typically has four babies to keep her salty form of mousehood sloshing along.

City mouse or country mouse. Rippling waters or waving grasslands. Rest assured that wherever we are in life, we can always find perky little ways to "eek" the most out of it.

The salt marsh harvest mouse (*Reithrodontomys raviventris*) is tiny, about the weight of a quarter, and it requires marshes for their deep cover (as shown here, in pickleweed). It lives only in the fragmented salt marshes along San Francisco Bay and its tributaries—wetlands that began disappearing after the California Gold Rush. The salt marsh harvest mouse is endangered.

California Newt

Walk your walk

California newts are the orange-bellied bridges between the worlds of water and dry land.

During the warm months, newts gather into animal burrows, deep leaf litter, or other moist nooks of the forest floor to *aestivate*, or undergo summer dormancy. During the wet months, they emerge from the forest and start walking. Their destination: the slow-moving creeks and ponds where they underwent their own metamorphosis. Here these amphibians will mingle and create new baby newts.

Many types of animal find their way back home using the stars (including the sun), magnetic fields, scents, landmarks, instinct, or even the guidance of their elders. California newts keep their migrations simple. They use their sense of body position, or *kinesthetic* sense, to march in straight lines back to the waters from which they came. Nevermind any rocks or objects that stand in their way: newts walk around them, then make corrections to maintain their linear course. These arrow-direct treks may be two miles long "as the crow flies." But not to small amphibians on the ground: these nimble journeys are just "as straight as the newt walks."

W arning: California newts (*Taricha torosa*) possess a toxin that, as far as we know, only a garter snake can safely injest. These newts live and walk in the California Coast Ranges and Sierra Nevada, where, during their migrations, they may amble *en masse* across busy roads. Tilden Regional Park in the Berkeley Hills protects its California newts by closing one of its roads to cars during wet months. In other areas of the world, amphibian tunnels and wildlife underpasses help save the lives of animals on the move.

Snow Leopard

Keep climbing higher

*P*aw prints fall fresh as snowflakes as a hefty male snow leopard climbs the hardscrabble of a mountain slope, the dark gray rosettes of his coat erasing the line between cat and rock. He reaches the treeless barrens above 17,000 feet and takes his seat upon his granite throne in stone-cold silence. Not one Himalayan snowcock will notice him, or whistle. Not one pebble will loosen under his paws, or go ringing down the precipice.

Fellow snow leopards climb steep broken slopes in China, Mongolia, India, Afghanistan, and eight other Central Asian countries. But vast landscapes separate the cats, making their occurrences as spotty as their coats. These kings and queens of the mountains journey through life on their own, eluding any need for a word, such as pride, to describe a snow leopard group.

The special times that snow leopards do share together are celebrated with sound: the cubs' soft mewing for mom and the hilltop wailing of a female to her suitor. Snow leopards also make happy "chuffing" sounds, their breathy version of purring. But these big cats usually stay quiet and lack the ability to roar. They mostly communicate through scents and clawed markings, long-distance messages that manage to survive where few things can.

Snow leopards finding these signs somehow understand, even mountains away: *We're connected.*

The hefty male cat climbs another hardscrabble rock face, his paws leaving momentary tracks in the snow. In the shelter of an overhang, he curls up onto his new stony throne, his luxurious tail becoming a scarf he wraps around his body and face. His eyes, like silver and sky, look down upon a herd of wild blue sheep … watching … waiting … to make a 30-foot pouncing leap and high-speed, rock-flying chase after the meal that will feed him for days. But for now, silver-sky eyes blink, as the Himalayan winds blow his fur scarf and carry his purring breath up into thin air.

Snow leopards (*Panthera uncia*) are secretive, medium-sized cats that weigh 80 to 120 pounds and command the inhopitable high-mountain terrain of Central Asia. Their patterned coats blend perfectly with their rugged surroundings; in fact, a snow leopard is practically invisible from a distance as it climbs a broken slope. Yet even a snow leopard cannot elude the threats to its species' survival: loss of natural habitat and prey, retribution killings by livestock keepers, poaching, persecution, and illegal trade. Snow leopards are endangered.

African Wild Dog

Give 'em something to wag about

A pack of African wild dogs is sleeping comfortably in the dirt. But one young male rises early. He's an underdog, who let the pups, elders, and alphas eat first during the night. Hunger may keep him from sleeping in, but it won't keep him from celebrating when the rest of his family wakes up.

Their morning ritual—their "greeting ceremony"—begins when the dogs rally together and break into a warble, softly, sounding more like gentle songbirds than a hunting pack. Everyone finds a face to lick, reason to leap, and something to wag about.

The celebration leaves all the dogs hungry, so the group gets into hunting formation: an orderly line of twenty wet noses behind windshield-wiper tails. The alpha pair then leads the pack out through the savanna, setting a loping pace that these dogs can maintain for hours. Wild dogs—"the fugitives of the bush"—are almost always moving, evading, searching for a new spot to spend the night, as the scent of their

kills can bring big trouble in the form of lions, hyenas, and leopards. So goes wild dog etiquette: it's perfectly polite to eat and run.

It is also alright to go searching for something more. And tonight, the young, sleep-deprived male dog stepped away from his usual place at the back of the hunting line. He trotted off in a new direction. There are risks in wandering the bush alone, but somewhere, maybe a few miles away, maybe one hundred miles away, a female wild dog (traveling with her sisters) has left her family pack and gone looking for him, too. When they meet, both of their tails will hit high speed.

This is more than puppy love. Tonight is the night to start a new pace and a new pack.

African wild dogs (*Lycaon pictus*), also known as painted dogs or African hunting dogs, weigh about half as much as spotted hyenas. Packs are family groups, and in large packs, both males and females will guard and feed pups born to the alpha pair. Group members may also offer help to the old and injured. Wild dogs have disappeared from 25 of their 39 home countries and are amongst Africa's most endangered carnivores.

Honeybee

Share your sweet gifts

*I*t all starts with sunshine.

The sun gives life to plants. The flowers give nectar to honeybees. And the bees share this nectar, mouth to mouth, with other workers in the hive. These worker bees are sisters, daughters of the queen, and as they pass the nectar around, water in it gradually evaporates; this makes the nectar grow sweeter. Finally, the bees store the leftover nectar in the honeycomb, passing it sweetly along, one last time, to the future.

*I*t all ends with sunshine.

One of life's last undertakings, the job of buzzing out to the sun-drenched flowers, is reserved for the elder honeybees, having already performed a progression of sisterly duties over their busy little lives. Prior to taking their flights,

these bees train in the hive, beating their wings like miniature fans. With this moving air, nectar in the honeycomb distills down even further, concentrating the flavor of the landscape: the brief, buttery bloom of tupelo trees along Florida's creeks, or the wind-blown lavenders of the British heaths, or the thin veil of flowers across the Middle East.

Honey is the gift of land and sky, kissed by the sisterhood of bees: sweet, sweet droplets of pure golden sunshine.

People have been honey hunting and beekeeping for thousands of years. Yet the honeybee's sweetest gift to humanity is its work as a pollinator. In the United States alone, honeybee pollination directly or indirectly supports one-third of the food people eat. Honeybees (genus *Apis*) are native to the Old World (Europe, Africa, and Asia) and have been introduced worldwide. Their colonies are collapsing in many parts of the globe, causing serious concern for food security and biodiversity.

Electric Eel

Be positive

Lightning issues from the mud as well as from the sky in the Amazon and Orinoco Rivers. Here, electric eels (ropy knifefish we call "eels") swim as slowly as they like through the darkness and murk, mighty and snaky like the rivers themselves, shooting 600-volt shocks at other fishes.

Electric eels have even meandered beyond their rivers and into the channels of human thought. Top scientists of the 18th century, including Benjamin Franklin, took up formal study of their awesome powers, and the "quiet" nature of the eels' shocks (and shocks from other electric fishes) ended up changing the very definition of electricity and the study of physics. We can also thank these eels every time we use a common battery. Italian physicist Alessandro Volta invented it in 1799 by building an imitation of an eel's electric organ.

Electric eels have a subtle side, too. They constantly pulse on a low "setting" equal to one-fiftieth their maximum shock power. This weak pulsing creates an electrical field around their bodies, from head to tail, positive end to negative end. As the eels swim, waves in their electrical fields bend around objects, helping them find their winding way.

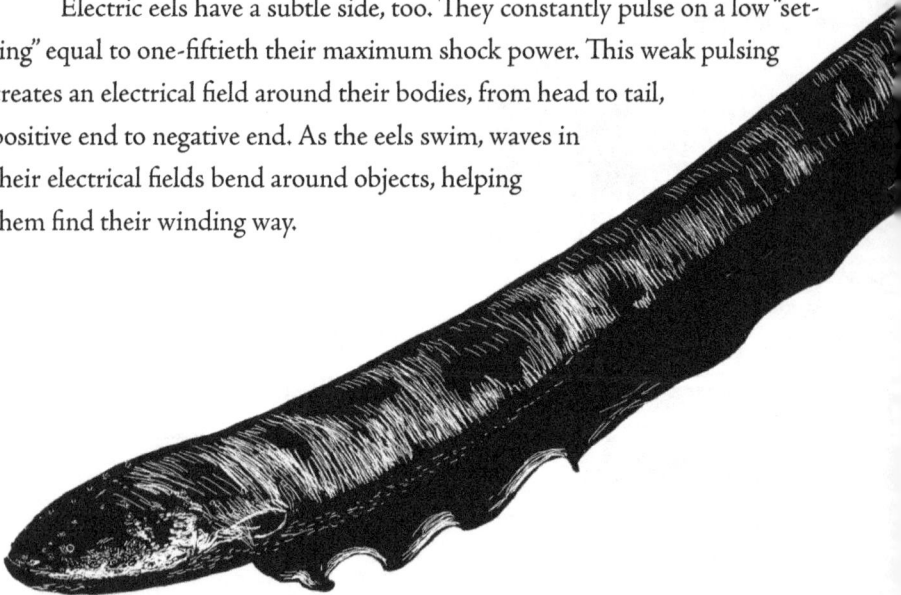

Weak pulsing also helps eels identify one another and attract mates, like electric songs. Scientists who have figured out how to hear the eels calling have crossed the boundaries between biology and physics. Such discrete sciences, however, are of our own design. Listening to the voice of electricity is truly a limitless science—the science of nature—in the language of living organisms and the power of lightning.

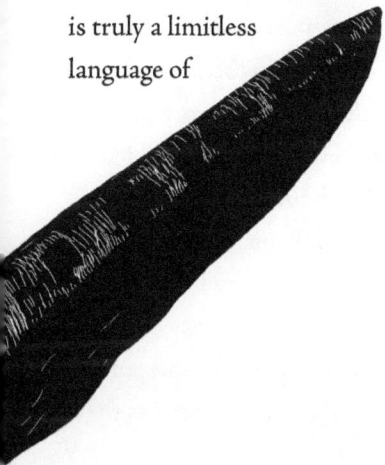

Electric eels (*Electrophorus electricus*) are top predators that wend their slender, six-foot bodies through major river systems of Brazil, French Guiana, Guyana, Peru, Suriname, and Venezuela. Males play an unusual role in reproduction: during the dry season, they build a foam nest with their saliva. A female will lay hundreds of eggs into the nest, and the male will guard the growing larvae until they disperse with the floods.

American Bison

Gather close—roam far

In springtime on the prairie, newborn bison pop up onto their feet like blooming flowers. They stand beside their mothers within minutes of entering the world, and soon, they go scampering, chasing, making mock charges at things they see in the grass. Playtime ends. And every calf, wide-eyed and woolly, returns to that plushy safe place between its mom's neck and beard.

Springtime is bonding time for mothers and calves. So the calves from last year, the yearlings, turn to one another for comfort. The boys rove in groups, although they tend to roughhouse. Already, the struggle for dominance is growing inside them. The yearling girls form a sisterhood that may last for life. One day they will raise their own wide-eyed, woolly babies together.

Here in America's Heartland, most original grassland lives belowground in the roots, in tangled miles that prevent the soil from washing away or blowing off on the prairie winds. Bison are a huge, snorting part of this land, and their cupped hooves shift, rather than compress, the precious soils as they walk, leg-chaps swaying side to side as they go.

Ten long feet of intimidation, horns protruding from a bonnet of wool, a roarer: this is the bison bull, patriarch of both prairie and herd. Few things can match his aggressive vitality or raw power. He can leap six feet up an embankment from a dead stand and outrun some racehorses. Yet he will drop his one-ton body to the ground, roll around on his arched back, then rise to his hooves through a pall of dust to roam once again with the herd.

An estimated 30 million bison once grazed the vast prairies, walking season after season, horizon to horizon, in great wild herds. Today, wherever these mighty animals are left to seek green grass and fresh water, every step of a hoof and sway of a chap reveals what is truly in their hearts ….

Home is always with your herd, wherever you may roam.

Also called American buffalo, American bison (*Bison bison*) comprise two subspecies: the plains bison, whose original range extended from northern Mexico to Alaska, and the wood bison of central Canada. Bison were overhunted and nearly exterminated during the 19th century. Since then, they have made a limited recovery. Small wild herds roam freely on public lands, like Yellowstone National Park, but the vast majority of bison living today are kept as commercial livestock. The American bison is a near-threatened species.

Zebra Swallowtail Butterfly

Let love land on you

Spring showers bring butterflies.

It all begins when the female zebra swallowtail butterfly goes flitting, batlike, through the forest. When she lands on a leaf, the male butterfly makes darting passes over her, sprinkling her antennae with his dust, *love dust*, made of tiny invisible scales that waft off his wings. Male-chemistry in the dust amplifies the chemistry between the two butterflies. And the she-butterfly shifts her attention from the wide leafy world to her suitor, literally showering his attention upon her.

Soon the female butterfly produces tiny eggs, like pale pearls of jade, and she hangs them under the leaves a pawpaw plant, a type of custard-apple. The pawpaw, in fact, serves as the "host plant" for zebra swallowtail caterpillars; it's all they eat at this stage of life. And eating is definitely their first love. These caterpillars will shed their skin six times to accommodate their bulging waistlines.

A caterpillar will eventually eat its fill and inch its way out under a leaftip, where it undergoes the most dramatic stage of its metamorphosis: it becomes a chrysalis. The animal's cocoon hangs from the leaf like an inverted question mark, while on the inside, its body breaks down into the soupy building blocks of transformation.

To what will it be reborn?

One month ago, the butterfly was a tiny jade pearl. Now this insect nymph of the air spreads her fantastic sword-tipped wings and embarks on her first darting flutter through the forest. She is a pollinator who carries pollen dust from flower to flower. But when she lands on a blackberry blossom, she feels that it is starting to sprinkle.

Could this be love?

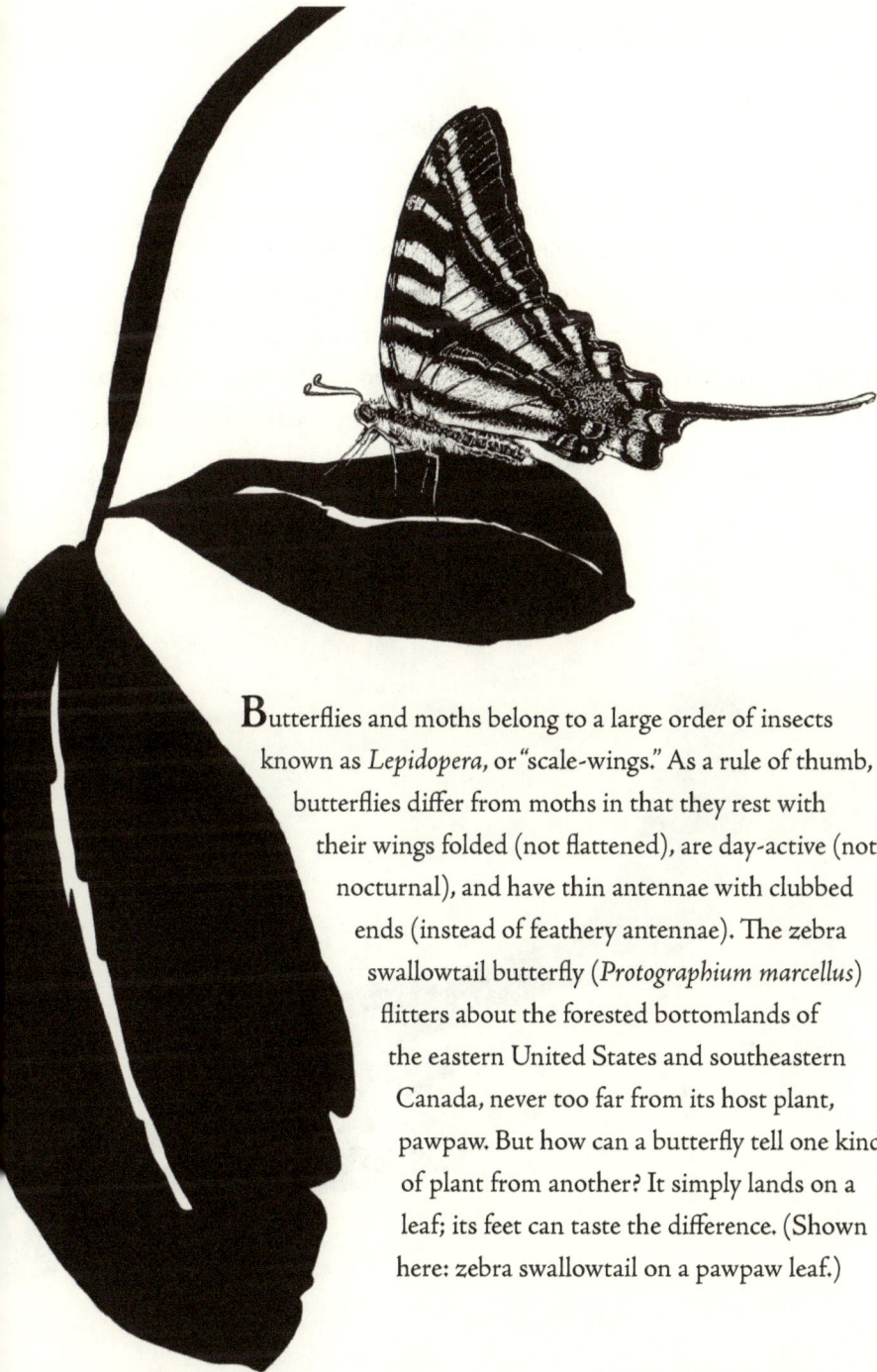

Butterflies and moths belong to a large order of insects known as *Lepidopera*, or "scale-wings." As a rule of thumb, butterflies differ from moths in that they rest with their wings folded (not flattened), are day-active (not nocturnal), and have thin antennae with clubbed ends (instead of feathery antennae). The zebra swallowtail butterfly (*Protographium marcellus*) flitters about the forested bottomlands of the eastern United States and southeastern Canada, never too far from its host plant, pawpaw. But how can a butterfly tell one kind of plant from another? It simply lands on a leaf; its feet can taste the difference. (Shown here: zebra swallowtail on a pawpaw leaf.)

Manta Ray

Use your edge

*N*othing is black or white in the world of the manta ray.

Mantas are fishes that swim like angels and look like devils, with those hornlike head fins and long, skinny tails. In fact, they are members of the devil ray family, whose pectoral fins, or "wings," can reach wider than 20 feet from tip to tip. And they swim, practically fly, through the ocean both day and night in an effort to sieve enough plankton from the water to maintain their weight— 2,800-pounds for the largest mantas.

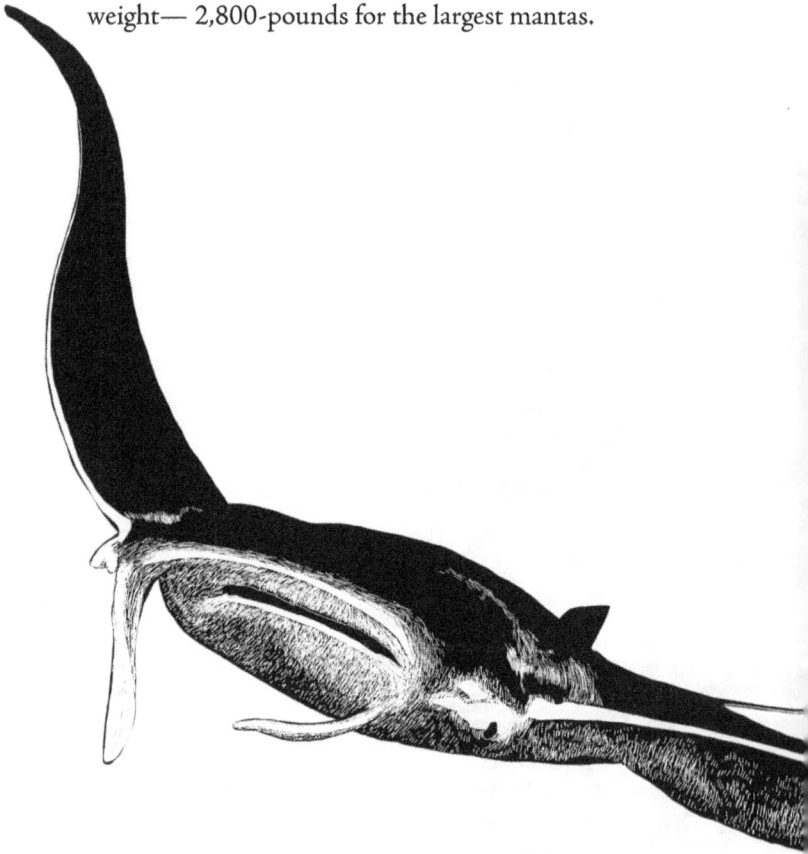

Yet even the ocean cannot seem to hold these massive fishes. Mantas may sail several yards out of the water into the great blue sky. Jumping may be a quick way to flee from danger, show off for a mate, or splash parasites off their rough, grainy skin. Their flat bodies slap back into the sea, making a crashing boom that resounds for miles. So just perhaps, a series of leaps is a form of communication, manta-style.

When it comes to big fishes, the rays' shark cousins hold top places on the scale. But flat out, gentle mantas are the grandest fishes that can take wing and fly.

Manta is the Spanish word for "blanket," and manta rays (genus *Manta*) are cartilaginous fishes, or elasmobranchs, like sharks, rays, and skates. The mantas' speed and open-water lifestyle give them no need for a stinger. They eat by filter-feeding and have only vestigial (remnant) teeth, about as sharp as sandpaper—leftovers from their ancestors that munched on shellfish. Two species of manta ray are recognized today, winging about the world's oceans as far north as New Jersey and as far south as New Zealand.

Palm Cockatoo

Drum up the support

Knock! Knock-knock-knock-knock! Knock-knock! Knock! Knock! Knock-knock-knock-knock! Knock-knock! Knock!

Loud, woody rapping in the tropical woodland is part drumbeat, part heartbeat. The male palm cockatoo—one of few birds that makes and uses tools—has clipped off a drumstick and is tapping out a rhythm on a tree trunk cavity, the doorway to his nest.

He only needs to sell one ticket to his show. And a female palm cockatoo is listening from the branches above, head cocked, crest raised, interested. She climbs down to investigate, quickly finding both the nest and her suitor quite attractive. Pair-bonding starts with a drumbeat—the beat that goes on for the rest of their lives.

Palm cockatoos (*Probosciger aterrimus*) are famous vocalists as well as drummers, giving a full repertoire of whistles and a call that sounds like the word, "hello." These large, smoky black members of the parrot family have striking red cheek patches—coloration that continues onto the tongue. Palm cockatoos live in far Northern Australia (Queensland) and the neighboring island of New Guinea, where old, hollowed-out trees provide nesting space, and rainforests offer sweet fruit.

Red Kangaroo

Put a little bounce in it

A baby red kangaroo gets his legs. He cascades out of his mother's pouch and begins hopping, wobbling, feeling out this sunburned land where even the dry bumps, clumps, and shrubs are new fascinating toys.

Excited, the young joey springs over to mom, hopping playfully around her hippie figure as she lies stretched out upon the red ground. With one big boing, he clears her kickstand tail. She gets up for a lighthearted tussle, briefly, before her joey wears himself out and falls into her like a saggy rubberband. Mom reaches out with slender hands, holds his face, and licks his ears until they are cool, wet, drooping.

Mom prefers evening activity, when her smoky blue fur blends into cooler sands. She is known as a "blue flier," fastest amongst the more than sixty types of roo. Her son, a "boomer," will eventually double her in size but will never match her bursts into highway speeds. At a moderate pace and over long distances, hopping is more efficient than four-legged running. It is also extremely rare in large animals, like red kangaroos. In the vastness and scratch of the Australian outback, kangaroos are the native grazers, pogo-ing versions of cows or sheep.

Breaking rains bring welcomed relief to this scorched land, but not today. So the young joey looks to his mother for shelter from the heat. He pokes his head inside her pouch and then springs into a somersault … he's in! Mom takes junior for a ride, as she leisurely scouts the red soils and notices, far in the distance, a refreshing glint of green.

Buckle up, Son! It's time to bounce!

The word, *kangaroo*, is one of the first English words borrowed from an Aboriginal language, as both Captain Cook and botanist Joseph Banks wrote it in their journals in 1770. (Banks wrote "kangaru.") Red kangaroos (*Macropus rufus*) are the largest living marsupials and range across much of the Australian Outback.

Barn Owl

Take a look around you

Barn owls are known as the "sweetheart owls" for their valentine faces. Yet, their courting behavior is equally dear. During the breeding season, the male woos his mate with romantic gifts—dead rodents—which she eats like a box of chocolates.

Heavier owl-hens lay more eggs, so the father-to-be takes on a new night job—that of feeding his mate. He may even offer too many treats, causing a furry, long-tailed surplus to build up in the nest. But better too much than not enough! The female sits "tightly" on her eggs for about a month, relying this whole time on her mate for his delivery service. Even after the chicks have hatched, dad silently goes about his sweet night's work, assuring that every hungry heart in the nest will be fed.

So take a look around: there's a lot of love in the air.

All owls are unable to move their eyes in their sockets, so they turn their heads to see the full view. The barn owl (*Tyto alba*) looks, in particular, for barn rafters, church steeples, tree cavities, and other safe, ready-made nooks where nest-building is not required. This owl is one of the world's most widespread birds—almost global—living near grasslands where it can hunt at night. By hearing alone, a barn owl can detect even the smallest vole moving or munching in the grass, then fly in for the kill.

Peccary

Wallow in happiness & root your friends on

*I*f it looks like a pig and walks like a pig, then it's not necessarily a pig. It could be a peccary!

Sharing the same family tree as pigs and hogs—but wallowing in the shade of its own branch—the peccary is native to the New World, where its three species are known by more than 200 local indigenous names. The word peccary most likely originated in the Tupí (native Brazilian) language, meaning "an animal which makes many paths through the woods."

Rainforest. Thorn forest. Desert. Peccaries aren't picky as long as they have one another. The herd disperses for a few morning hours, to forage, their reassembly a practice in baritone grunting noises. Some peccaries greet nose-to-nose. Others stand flank-to-flank, rubbing in one another's musk. Then everyone follows a path back to the bedding sites: cool, comfy bowls hollowed out of the ground by their long, disklike noses. Some peccaries take a soothing wallow before they go to bed, cuddle their stiff-backed bodies into one another, and purr themselves off to sleep.

By the afternoon, peccaries are perky again. Babies stand at attention under their moms, ready for the next adventure down the trail. But before the herd scuttles off, a few more noses nuzzle a few more friendly necks. These sensitive snouts will also sniff and dig out plant-foods underground, even in the driest of times. This rooting by peccaries tills the soil, preparing it for new seeds—also dispersed by peccaries.

And so the peccary is not just an animal that makes many paths through the woods. This animal also helps make the woods.

Peccaries (family *Tayassuidae*) live in the Americas from Arizona to Argentina. Shown here is the collared peccary (*Pecari tajacu*), also known as the javelina. It is the most widespread and adaptable peccary, and it can root out a living in suburbs. The remote yet gregarious white-lipped peccary (*Tayassu pecari*) is considered a near-threatened species due to overhunting and loss of neotropical habitat. The Chacoan peccary (*Catagonus wagneri*) is endangered and was once known to science only through its remains.

Greater Flamingo

Keep your balance

*I*t grunts. It moves. And it's downy gray. It looks just like an island. But this "island" is a huddled mass of hundreds, maybe thousands, of greater flamingo chicks. It's called a crèche. And it's hungry!

Bearing food, adult flamingos fly in toward the chicks, whose rattling calls sound more like a factory of squeak-toys than a gathering of soft baby birds. Above the din, a chick manages to recognize its parents' gooselike honks—its dinnerbell! Then knobby little legs come running.

A good long life awaits these young birds, perhaps sixty years, with a few males reaching the height of a man. Some chicks will grow to adulthood near the precious wetlands of home, in Africa, Asia, and the rim of the Mediterranean Sea. Others will move on to new shallow waters, helping maintain the delicate balance of pink and black feathers around the world. Northern flamingos will disperse south. Southern flamingos will disperse north. And those in the center will fly in any marshy direction filled with the sound of curved-billed happy horns.

The flock finally quiets down for rest time, when these stick-figure birds balance on one leg in the water. Many have set out to explain this one-legged form of relaxation, but only the flamingos know the real reason why. So, like the flamingo, we perform balancing acts of our own. Sometimes, we can uncover the answers to life's mysteries. Other times, we can learn to accept what "just is." Yet oftentimes, the light emerges with one long and willowy step back.

Flamingos are ancient birds whose neck and legs are longer, relative to body size, than any other type of bird. They also have a uniquely shaped bill, which they use for filter-feeding in similar fashion to a baleen whale. The greater flamingo (*Phoenicopterus roseus*) is the tallest, largest, and most widespread member of its rosy feathered family, its flocks gathering broadly across the Old World in wetlands near arid lands.

Frilled Lizard

Stand and deliver

Real dragons don't breathe fire. Nor do they launch winged attacks. Real dragons, the frilled lizards of Australia and Papua New Guinea, turn to face their threat and flare up into a big, hissing frill. And whether that mouth is pink or yellow, it is always a red alert.

Yes, real dragons just might outmatch the dragons of legend. For their bravery is all in the bluff.

The frilled lizard (*Chlamydosaurus kingii*) is one of the most recognizable dragons, or member of the reptile family *Agamidae*, which includes such species as the Chinese water dragon, bearded dragon, and slow-moving thorny devil. Dragons generally flee from danger, and the hefty frilled lizard can sprint on two legs and run up trees. Or it may freeze and open its mouth, which automatically raises the alarmingly bright ruff around its neck.

Desert Bighorn Sheep

Rock steady

A baby ram is born high upon a stone ledge, no bigger than his mother's body lying down. Spread out below them is the rugged floor of the Mojave Desert, its stones and boulders filling the expanse to the next mountain range. And here, balanced on the edge of the world, the newborn lamb will press little hooves to rock, step away from the drop-off, and go gamboling after mom to reunite with the herd. In the steep and rocky world of the desert bighorn sheep, a lamb's first steps are the most precarious.

By two years of age, this young sheep leaves his mom's group of other mothers, grandmothers, and lambs to go travel these dry, crumbling mountains with the rams. To him, the same huge curled horns that can rip into barrel cacti for water and clack together in great clashes for dominance also signal something greater: wisdom. He and other young rams follow their big-horned elders, walking sure-footedly in their hoof steps, just as they walked in the hoof steps of those who came before them. By the autumn rut, or mating season, the rams make tracks back to the company of the female herd, creating the year's largest gathering of the desert bighorns.

The next steady generation of bighorn sheep soon comes along, and then the next generation, along with bighorn traditions. Ewes will pass down the knowledge of grassy pockets and safe passages to their daughters; big rams will pass this knowledge down to smaller rams. Their migratory traditions are as solid as the desert's warm sandstones and cool granites that, for millennia, have born Native American carvings of bighorn sheep. To live in the stone—through both imprints and footprints—is what it truly means to be rock steady.

Desert bighorn sheep (*Ovis canadensis nelsoni*) have special soft-bottomed hooves for gripping rough terrain. These sheep graze in the arid ranges of the American Southwest and northern Mexico, never too far from a steep or rocky hillside, where they can make a surefooted escape. The peninsular bighorn sheep (a population segment of the desert bighorn) is endangered due to habitat loss, disease from domestic livestock, and a history of indiscriminate hunting.

Chambered Nautilus

Let inspiration spiral around you

The swirling nautilus shell holds more than the animal's soft body; it holds a legacy that dates back 500 million years. Chambered nautiluses are the only living descendents in their direct family line which is older than fish, squid, or even ammonites, the nautilus's extinct cousins we know as fossils.

A growing nautilus slowly builds out its shell as a coil of about thirty chambers, each one much more cavernous than the next. This particular coiling shape follows a logarithmic spiral (a special, fast-expanding curve), and it is one of nature's perfect designs, like the seed pattern of a sunflower, the swirling of a whirlpool, or the starry arms of a galaxy. The animal resides in the shell's last and biggest chamber. Most other chambers contain gases, enabling a heavy two-pound nautilus to become featherweight underwater and ready to swim on.

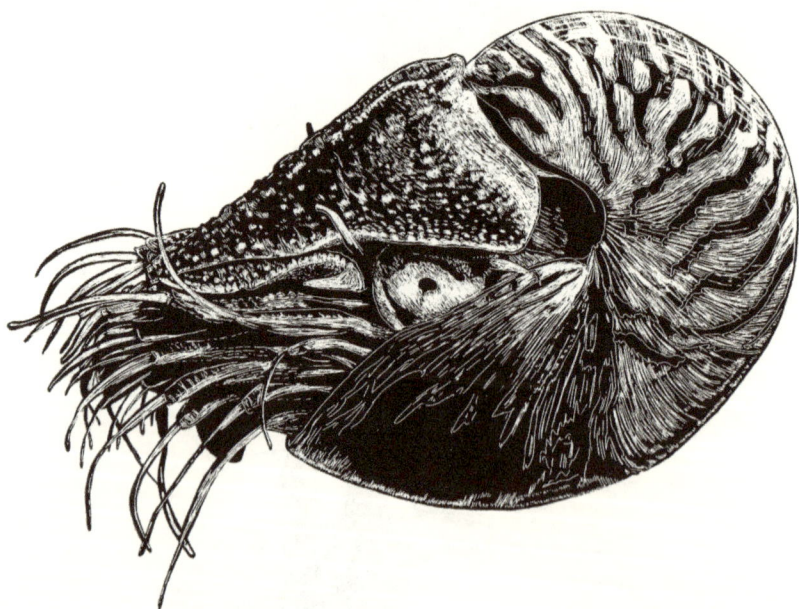

In fact, the nautilus spends its day swimming—jet-propelling—through waters as deep as 1,000 feet, sometimes much deeper, as long as its shell can withstand the crushing pressure. From the dark abyss, the animal somehow knows when the sun has set. Then it rises, swiveling as it goes, to feed along coral reef walls in shallower water. By daybreak, the nautilus returns to the depths and the cool, calm security of home—a pattern in life made possible by the sturdy yet harmonious architecture of its shell.

So pleasing is the nautilus's shape that similar spirals can be found far beyond the sea, in such inspired works as cave-wall carvings and the scrolls atop Greek columns. Perhaps the nautilus will continue to spin its magic for another half a billion years. And everything will turn out, and spiral out, beautifully.

Chambered nautiluses (family *Nautilidae*), or pearly nautiluses, live in tropical waters of the Indian and southwestern Pacific Oceans. They are cephalopod mollusks, like octopus, squid, and cuttlefish, with as many as ninety tentacles. Nautiluses are the only living cephalopods, however, whose shell remains on the outside, rather than the inside, of the body.

American Beaver

Still the waters

*T*he land is unlocking. Winter's hold is melting away. And the promise of spring trickles, rushes, swishes through the streams, whispering special instructions to the beavers: *It's time to build.*

A pair of beavers sets the foundation for their dam first, locking a row of poles across the stream channel. Then they stack their woodpile, facing it upstream and headlong into the poles—the beaver way of quieting that itchy sound of running water which stimulates their need to do woodworking.

Upstream from the dam, water quietly swallows its banks, giving the beavers their pond … and more. The pond is their superhighway, making distant forest a quick swim away. It's their refrigerator, where they float bark-covered logs for good eating later. It's their escape hatch, too cold and wet for wolves and black bears. It's even their garden, bursting with delectable yellow pond lilies and water shield.

Soon the hum of dragonfly wings rises from these still waters, where American black ducks join the feathered many that come to nest, wade, waddle, and sing. The thuddy ribbet of the green frog and long, tweeting trill of the American toad give way to the sounds of summer, when beaver kits slap the pond's surface with their baby tails and purr to one another in their lodge. Every now and then, the peaceful pond echoes with the trickle of water seeping in small cascades through the dam. Not for long, though, after its whisper reaches the ears of the beavers.

Beavers (genus *Castor*) take their Latin name from a complex fluid they produce, *castoreum*, which people have valued for thousands of years as a medicine and fragrance. The American beaver (*Castor canadensis*) weighs up to sixty pounds and is a rodent native to the forested waterways of the United States, Canada, and Mexico.

Spider

Suspend your disbelief

"OSpider, how do you infuse the world's finest silk with twice the strength of steel?"

"My dear friend," replied Spider. "Explanations are of little matter. The answer lies in *believing* that you can hang your dreams on one perfectly spun thread."

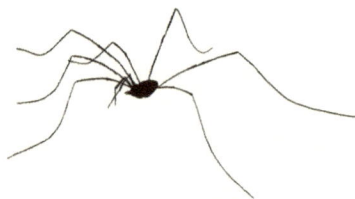

Spiders (order *Araneae*) thrive in tropical areas, but over the course of 300 million years, have perfected survival in almost every habitat type, from deserts to pond water. Some may even go airborne when migrating to new areas, "ballooning" on long filaments of silk drawn out upon the winds. Tiny airborne spiders, often spiderlings, have been spotted thousands of feet in the sky and by ships out at sea. Shown on the right is a male red-backed jumping spider (*Phidippus johnsoni*) whose real-life length is less than half an inch.

Loggerhead Sea Turtle

Reach for an ocean of dreams

A loggerhead sea turtle pulls her heart-shaped shell out of surf, drags her 250-pound body up the beach, and digs a pit with her flippers. She is one of nature's grandest gardeners who plants her eggs, like seeds, into the warm, moist sand. With a few sandy flicks, she covers up her eggs then returns to the sea. Her job is done. And Mother Earth, instead of the mother turtle, will keep these eggs safe and warm.

Loggerhead eggs in the middle of their crowded nest stay warmer; they become the females. Eggs lying closer to the sand go slightly cooler; they become the males. All hatch after about two months, the movement of the first active turtles stimulating their brothers and sisters to break free from their shells.

By nighttime, one hundred baby turtles will come flippering out of the same nest, then they dash toward the brightest horizon—the ocean—glowing in the light of the moon and stars. Their gauntlet takes them past crab claws and stabbing beaks to freedom in the surf. For these tiny turtles weren't born to run; they were born to the sea.

In fact, the whole ocean is a turtle dreamworld. Loggerheads from Florida's beaches will eventually swim a giant lap around the North Atlantic, using magnetic fields to find their way. Others may paddle 7,500 miles across the Pacific Ocean—a journey that can take them six years. And hopefully one day, maybe in a decade, maybe in two, another female loggerhead will cross countries and continents to return home, to the beach where she hatched, to lay eggs of her own. The end of the sand is just the beginning of the possibilities.

Loggerheads (*Caretta caretta*) are circumglobal sea turtles named for their large heads with shell-crushing jaws. Young loggerheads live in deep water, hiding in drifting mats of brown algae, called *Sargassum*, and feeding on soft food, such as jellies and even the poisonous Portuguese man-of-war. Years later, halfgrown loggerheads move to the shallower coast and prey on crabs, clams, and other crunchies. Sea turtles, in general, face serious threats both on their nesting beaches and at sea. The loggerhead sea turtle is internationally classified as endangered.

Red-Crowned Crane

Make life a dance

*R*ed-crowned cranes turn wetlands into dancefloors. Pairs of these great snowy birds flap, bounce, spring as high as ten feet in the air. Some cranes pick the grass and toss it like confetti during their honkfest of dance. In the quieter moments, males and females bow to one another, their chests out, wings back, and long graceful necks drawing the outline of a lover's knot. *We are bonded.*

By springtime, cranes do a different kind of dance, as mom and dad arrange to take turns warming eggs in the nest. They also give trumpeting calls out to their chicks a few days before they hatch; the chicks peep soft reponses from inside their shells. Crane chicks, called *colts* for their long legs and knobby knees, stand up and follow their moms within hours of emerging upon the marsh. In true crane style, they can also flap tiny downy wings and dance!

If red-crowned cranes needed a reason to do their fancy moves, perhaps it would be life itself. As chicks or adults. In groups or alone. For bonding, motor development, or just a release of excitement, these stately birds readily trip and twirl between the Earth and sky.

It is said in Japan that the folding of one thousand origami cranes will make one's wish come true. No wonder, with wings wide open and ready to dance.

When it is not dancing, the red-crown crane (*Grus japonensis*) stands about five feet tall and is one of the world's largest cranes. Populations in Japan stay in residence yearround, whereas their counterparts on the Asian mainland of East Asia migrate between breeding and wintering areas. These birds are endangered, owing largely to loss of continental wetlands.

Pearl Oyster

... and let the world be yours

The pearl oyster glues itself to a rock or reef, and manages, in its own still way, to create beads of light in the sea. It's beautiful when two worlds come together—one animal, the other gem—between two plain-looking shells.

Sheen on the inside of the oyster's shell and a pearl's glossy luster come from *nacre*, or mother-of-pearl, which actually comprises the same substance as "regular" shell, except in crystal form. The oyster makes a pearl by laying down these microscopic crystals in thin sheets, like the layers of an onion around a solid core. Other organic layers lend delicate color to the pearl, from white to a pale rainbow of hues to dark, mirrorlike black. A trace amount of the water in which the oyster lived also stays within the pearl, making every pearl from every oyster its own gleaming tribute to the sea.

Pearl-making seldom, if ever, begins with a grain of sand. But it does start when an oyster turns a negative—an injury, a tiny parasite, or another intrusion—into a positive. So when you feel the whole world is yours, then anything is possible, even something as precious as a gem, simply by being your own luminous self.

Oysters are not the only mollusks that produce pearls. Some of the rarest pearls come from abalones, and the queen conch forms exquisite beads covered in flame patterns instead of mother-of-pearl. Pearl oysters (genera *Pinctada* and *Pteria*) are edible to strong-jawed fishes, like sharks, eels, and rays, but not usually to people. Instead, these animals have done a beautiful, global job of turning tropical oceans and seas into gleaming treasure chests.

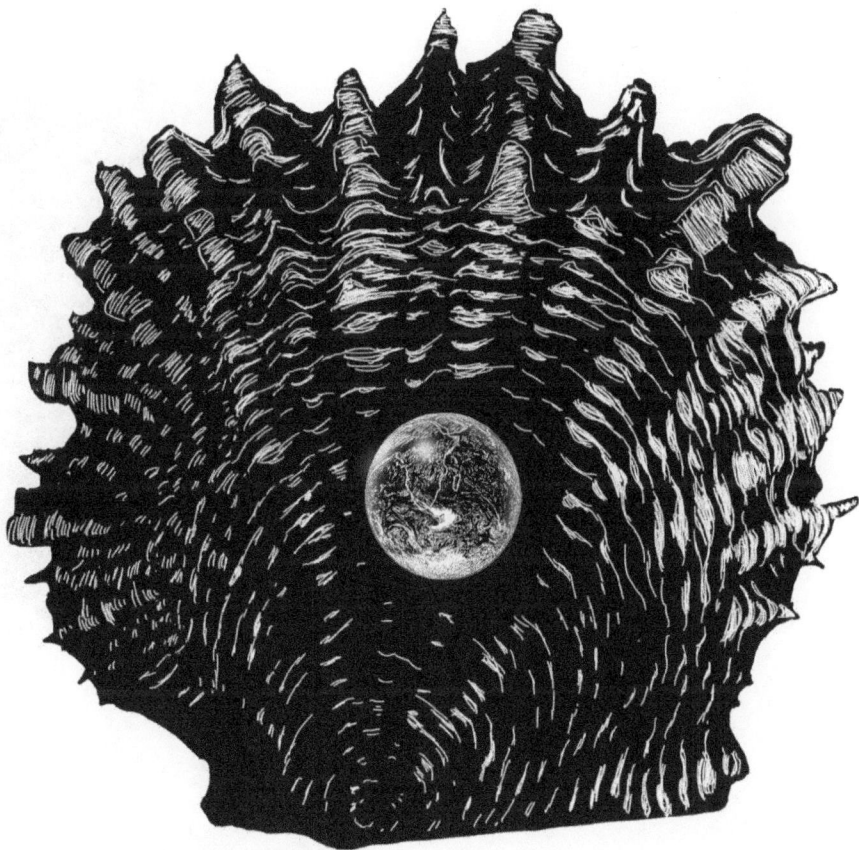

www.ingramcontent.com/pod-product-compliance
Lightning Source LLC
Chambersburg PA
CBHW021414090426
42742CB00009B/1146